O9-AIG-759

First published in Switzerland under the title *Kennst Du den Unterschied?*
by Michael Neugebauer Verlag AG, Gossau Zürich, Switzerland.

First published in the United States, Canada, Great Britain, Australia, and New Zealand in 1995
by North-South Books, an imprint of Nord-Süd Verlag AG, Gossau Zürich, Switzerland.
First paperback edition published in 1997.
Distributed in the United States by North-South Books Inc., New York.

Library of Congress Cataloging-in-Publication Data
Bischhoff-Miersch, Andrea.
[Kennst Du den Unterschied? English]
Do you know the difference? / text by Andrea and Michael Bischhoff-Miersch ; paintings
by Christine Faltermayr ; [translated by Rosemary Lanning].
"A Michael Neugebauer book."
1. Animals—Miscellanea—Juvenile literature. [1. Animals—Miscellanea.]
I. Bischhoff-Miersch, Michael. II. Faltermayr, Christine, ill. III. Title.
QL49.B61513 1995 596—dc20 94-40035

A CIP catalogue record for this book is available from The British Library.

ISBN 1-55858-371-8 (trade binding) 10 9 8 7 6 5 4 3 2
ISBN 1-55858-372-6 (library binding) 10 9 8 7 6 5 4 3 2
ISBN 1-55858-699-7 (paperback) 10 9 8 7 6 5 4 3 2 1
Printed in Belgium

For more information about our books, and the authors and artists
who create them, visit our web site: http://www.northsouth.com

Text by Andrea and
Michael Bischhoff-Miersch
Paintings by Christine Faltermayr

Do You Know the Difference?

Do You Know the Difference?

A MICHAEL NEUGEBAUER BOOK
NORTH-SOUTH BOOKS / NEW YORK / LONDON

GORILLA

Female gorillas look very similar to chimpanzees.

Gorillas are the largest and strongest of the anthropoid apes, the group of primates most closely related to people. A fully grown male gorilla can be as tall as six and a half feet (200 cm) and as heavy as 660 pounds (300 kg). Though powerfully built, gorillas are gentle creatures, much more placid than chimpanzees. Stories of fierce gorillas killing or abducting people are not based on solid evidence.

The picture shows a female lowland gorilla. If you see a gorilla in the zoo, it is more likely to be a lowland gorilla than the much rarer mountain gorilla, which has longer, shaggier hair. Both groups live in Africa, as do chimpanzees. Gorillas eat only plants. They live in family groups, usually led by a silverback, an older male with silver-grey hair on his back. All anthropoid apes are endangered species because of the continuing destruction of the tropical rain forests.

This profile shows the massive size of a male gorilla's skull.

CHIMPANZEE

Some chimpanzees have pale faces, like this female.

The chimpanzee's ears are large and prominent.

Chimpanzees are our nearest relatives in the animal kingdom. They are very intelligent, and unlike gorillas, they sometimes use simple tools. For example, they will push a stick or stem into a hole to fish out insects. The older animals show the young ones how to do this. Young chimpanzees stay with their mothers for ten years—much longer than other animals. Chimpanzees are smaller than gorillas. Even so, a young male is as strong as three humans. Though their diet is mainly fruit, they sometimes kill and eat small animals. In addition to gorillas and chimpanzees, there are two other species of anthropoid apes: gibbons and orangutans. Both live in Southeast Asia.

Note the small ears, the rounded back, and the knobby forehead of this cow elephant. Her tusks are too small to be seen.

Most people think that the only way to tell the difference between Indian and African elephants is to look at the size of their ears. The Indian elephant certainly does have smaller ears than the African, but if you look carefully you can find a number of other differences. The Indian elephant is smaller than the African, but because of its sturdier build it looks fatter. Its spine is arched, and the middle of its back is the highest part of its body. The forehead of the Indian elephant has two small humps. Only the bulls have large, visible tusks. The tusks of a cow are so small that they are usually completely hidden by her upper lip.

In Asia, some elephants are trained to work. Usually they help with logging, dragging away or carrying the felled trees.

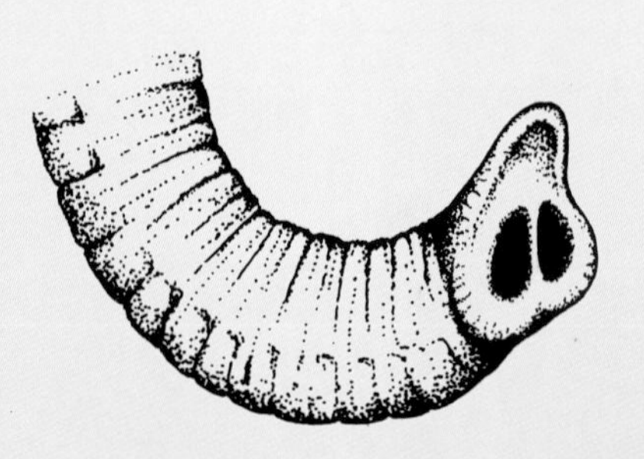

At the tip of the Indian elephant's trunk there is a single lip.

AFRICAN ELEPHANT

You can recognize the African elephant by its big ears.

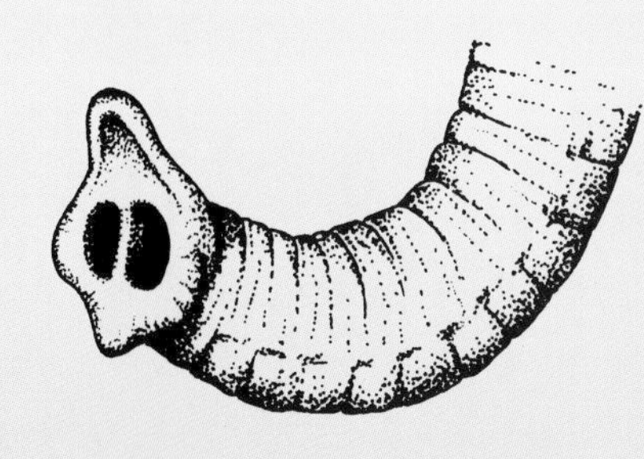

The tip of the African elephant's trunk has two lips.

An African elephant can grow as tall as 10 feet, 10 inches (330 cm) tall at the highest point of its body, which is the rear section of its back. It has a smooth forehead, looks slimmer than the Indian elephant, and has longer legs. Both bulls and cows have large tusks, which they use to dig for water in the dry season.

African elephants live on the savannah, a hot, dry grassland with few trees for shade. Their ears, which are much larger than those of the Indian elephant, are held out to make them look more frightening when they threaten or charge an enemy, but they have other uses too. The elephants flap their ears to drive away flies and to fan themselves. The skin on the inner side of the ear is thin, and there are large blood vessels just below the surface, to help the elephant cool down quickly as it fans currents of cool air.

The great white shark can grow to more than 19 feet (600 cm) long. It eats fish, sea birds, and seals.

Sharks are fish. Like stingrays, they belong to a group called cartilaginous fish, with flexible, gristly skeletons instead of hard, bony ones. Sharks have existed on Earth for 100 million years—twice as long as dolphins. Like all fish, sharks are cold-blooded. Their body temperature is about the same as the water in which they live. They do not need air to breathe, because they take oxygen from water passing through their gills. Like a dolphin, a shark propels itself with powerful strokes of its tail, but because its tail fin is vertical—going straight up and down—it must swish its tail from side to side and can only swim forward.

Sharks are not as dangerous as many people think. Worldwide, only about 100 people are attacked by sharks each year. Compared with the millions of people who are hurt or killed on the roads annually, this is a very small number.

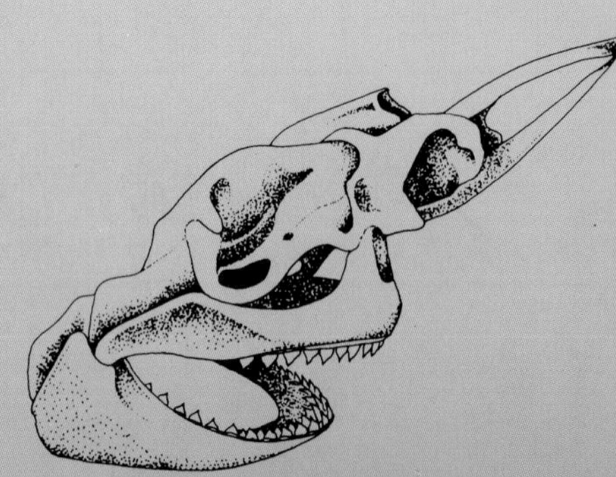

You can see from this skull that sharks have more than one row of teeth in each jaw. When a tooth breaks off, the one behind moves forward to replace it.

The common dolphin, shown here, is the best-known member of the dolphin family.

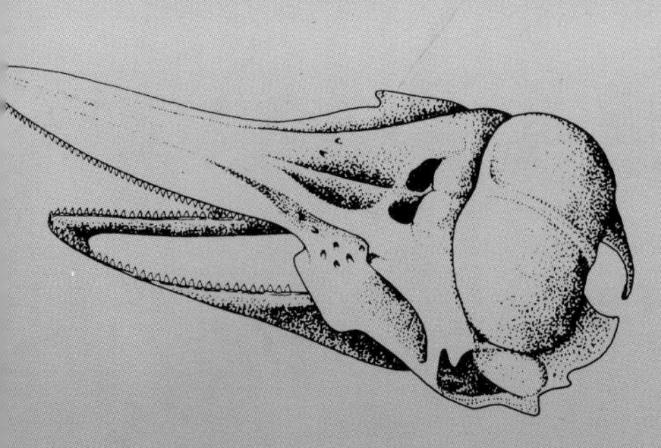

At the top of the skull you can see the blowhole, through which the dolphin breathes.

Dolphins are small whales. Like horses and dogs, they are mammals, which means they suckle their young, they are warm-blooded, and their babies develop in the mother's womb. Like us, dolphins breathe with lungs. That is why they have to keep coming to the surface for air.
Sharks and dolphins are only distantly related to each other. They look similar because the body of each animal is perfectly adapted for living and moving in the sea. The dolphin is the better swimmer, however, because its tail fin is horizontal—straight across. By pushing its tail up and down, it can swim in any direction it wants, just like a person wearing swim fins.

Dromedaries are thinner than Bactrian camels and have shorter hair.

DROMEDARY, OR ARABIAN CAMEL

Dromedaries have only one hump. They are thinner and have longer legs than Bactrian camels. Their humps are stores of fat, which provide the animal with reserves of energy if it cannot find food. Dromedaries that have had no food for a long time temporarily lose their hump. Dromedaries live in North Africa and Southwest Asia. They have been kept as domestic animals for more than 4,000 years. No other large animal can live without water for as long as a dromedary. They can survive without drinking for weeks. Human beings die if they lose a little over a tenth of their body weight through dehydration. A dromedary can lose 40 percent, nearly half of its body fluid, and then replace it with a single, long drink.

Camel riders in North Africa wear turbans to protect themselves from the sun, sand, and dust.

BACTRIAN CAMEL

Bactrian camels have a thick, woolly coat that grows even thicker in winter.

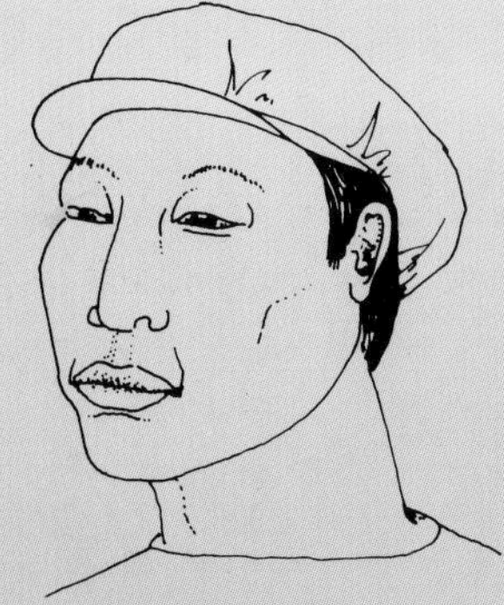

Mongolian shepherds use two-humped camels for riding and as pack animals.

The two-humped camel is also called a Bactrian camel, to distinguish it from other members of the camel family, such as dromedaries and llamas.
The Bactrian camel and the dromedary are closely related and can crossbreed. Their offspring have only one extra-tall hump. Apart from the number of humps, there are other differences between dromedaries and Bactrian camels.
The Bactrian camel lives in central Asia, where it needs its thicker, woollier coat to protect it from the cold. It also has shorter legs and a stockier build.

The spotted hyena is bigger than the African wild dog and about twice as heavy.

Although hyenas look like dogs, they are not closely related to them. The cat family and the little mongoose are much closer relatives. With a broad head and powerful jaws, the hyena is more thickset than a wild dog, and has a sloping back. The spotted, or laughing, hyena shown here often lives alongside the African wild dog. Both are hunters. It was once thought that hyenas were cowardly and fed only on the remains of animals killed by larger predators such as lions. This is quite wrong. In the Ngorongoro Crater in Africa, researchers have observed that hyenas often hunt for prey, only to have it stolen from them by lions.

Hyenas have powerful jaws that can crush big bones.

AFRICAN WILD DOG

African wild dogs have an irregular pattern of patches on their coats, and each animal has different markings. They look more elegant than hyenas.

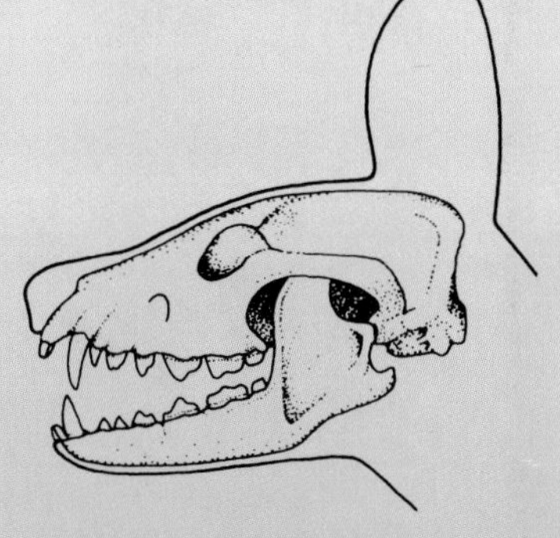

The jaws of a wild dog are very similar to those of a wolf or a domesticated dog.

African wild dogs are also known as Cape hunting dogs, and although the pattern on their coats may remind you of hyenas, they are more closely related to wolves.
Like wolves, they hunt in packs, working as a well-organized team. While one chases a gazelle, zebra, or wildebeest, the others cut off its escape. These dogs are very supportive and affectionate towards each other. If weak members of their group have no success in hunting, the others give up part of their own kill.
African wild dogs have become rare. This is mainly because farmers have relentlessly hunted them in the past, seeing them as a threat to their cattle. Many have also been killed by traffic on the roads.

This is the common zebra, also known as Burchell's zebra.

Zebras belong to the same family as horses and donkeys. Apart from their stripes, the only thing zebras have in common with okapis is that both live in Africa. But zebras prefer the wide, grassy plains, and avoid the forests. Unlike giraffes, zebras are not solitary animals, but live in groups. Common, or Burchell's, zebras sometimes congregate in enormous herds, and take part in mass migrations with wildebeest and gazelles. People have always wondered why zebras have stripes. Researchers now believe that these markings protect them from disease-carrying insects such as the tsetse fly. Their complex, many-faceted eyes do not recognize the zebras as solid shapes, because the stripes break up their outlines. The pattern of stripes on each zebra is unique. The animals recognize each other by these individual patterns, just as we recognize faces.

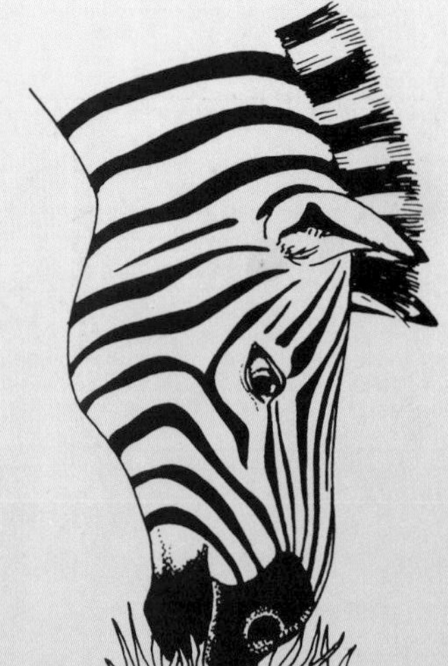

Like all horses, the zebra eats mainly grass.

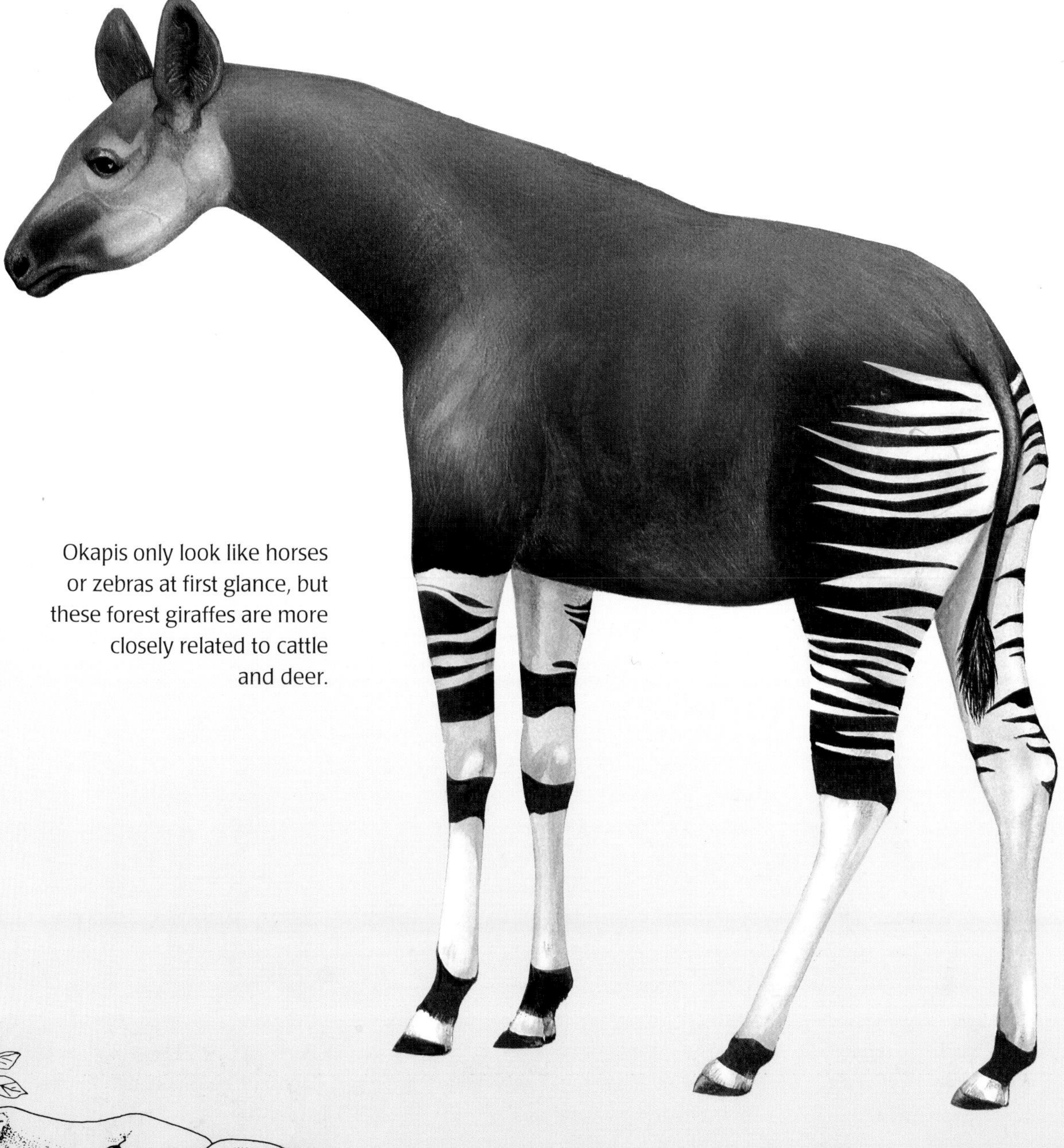

Okapis only look like horses or zebras at first glance, but these forest giraffes are more closely related to cattle and deer.

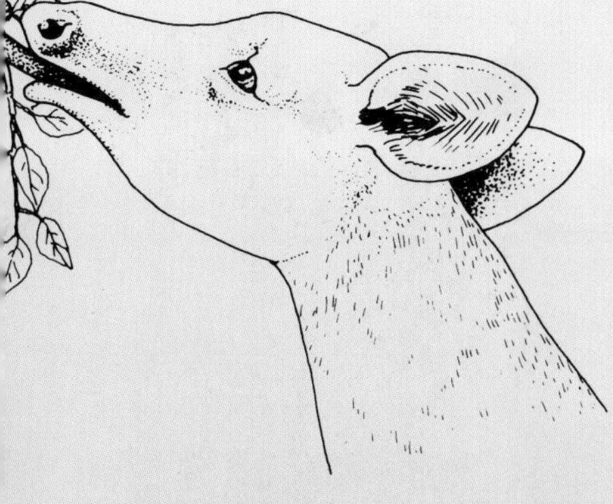

The okapi has a long tongue, which it uses to tear off leaves.

The okapi was the last large animal to be discovered by European explorers in Africa. In 1900 an English scientist sent an okapi skin back to London from the tropical forest near the Congo River. Misled by the striped legs, scientists immediately thought they had found a new type of zebra, but the okapi is not closely related to the zebra. When the skull arrived in London six months later, it was clear to the experts that the unnamed animal should be classified as a giraffe, despite the fact that the okapi's skin has a different pattern from the giraffe's. The okapi is also a much smaller and stockier animal. This is because it lives in the rain forest, where it wanders alone through the undergrowth and, like other giraffes, feeds mainly on leaves.

American alligators can reach a length of up to 12 feet (360 cm).

Alligators and crocodiles belong to an ancient order of animals that existed as long as 180 million years ago—even before their relatives the dinosaurs. The best-known species of alligator lives in the southern United States. It is called the American alligator (*Alligator mississippiensis*) and has a broad snout that curves upwards slightly at the end. These huge reptiles are quite common in the Florida swamps, where they have shaped the landscape by digging water holes to help them survive the summer droughts. Trees and bushes take root in the mounds of earth thrown up on either side of the water holes. Except for the smaller Chinese alligator of the Yangtse river and the even smaller caimans of South America, alligators live in North America.

When the alligator's mouth is closed, none of its bottom teeth are visible.

CROCODILE

Crocodiles look slimmer than alligators and have narrower snouts. Nile crocodiles can grow up to 21 feet (650 cm) long.

Some of the crocodile's lower teeth can be seen even when its mouth is closed.

Crocodiles are closely related to alligators, but have a longer, narrower snout. There are thirteen different types of crocodile living in the tropical areas of the world. Nile crocodiles, like the one shown here, could once be found in rivers and lakes all over Africa, but at the beginning of the twentieth century they disappeared from many places because they were hunted for their skins. Since crocodile hunting has been banned, the number of Nile crocodiles has increased again. Crocodile skin is still in strong demand, but more and more of it comes from farms, where the crocodiles are specially raised for this purpose. At least wild crocodiles are less threatened now.

To the ancient Egyptians crocodiles were sacred animals. They kept them in golden ponds and preserved them as mummies when they died.

The jaguar is heavily built and powerful.

Jaguars live mainly in the rain forests of South America. They are much heavier and sturdier than leopards and have a noticeably larger head. Although in other ways they look very similar to leopards, their lifestyle is more like that of the tiger. Unlike leopards, jaguars spend more time on the ground than in the trees, and are quite at home in water. With its powerful jaws, a jaguar can crack the skull of a wild pig as if it were a nutshell. The jaguar's roar sounds like the deep, hoarse bark of a big dog and is very similar to the leopard's.

The rosette patterns on the jaguar's coat are bigger and have a central spot.

LEOPARD

Leopards are slimmer than jaguars.

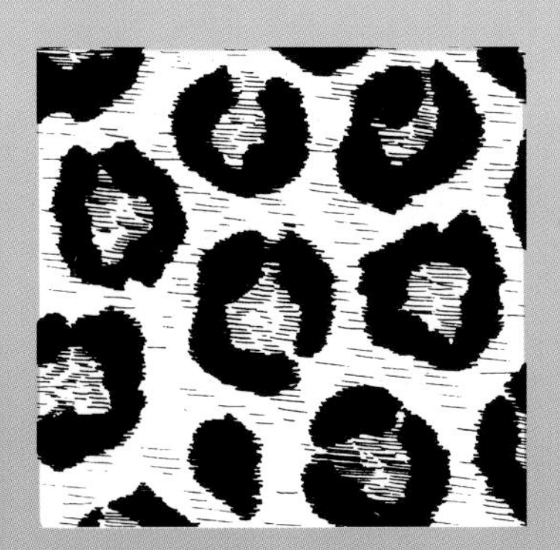

The leopard's yellowish fur is patterned with rosettes with no central spot.

The leopard is slimmer than the jaguar. It weighs only half as much, although it is roughly the same length and the same height. Its chest is narrower, and its legs and tail are longer. Leopards are found on two continents, Africa and Asia. They are very adaptable and can live in steamy jungles, dry, grassy plains, hot deserts, and even the snowy forests of eastern Siberia. The thickness of their coats and the markings on it can vary widely, according to the area in which these big cats live. Not all leopards are spotted. Some are completely black, and are called black panthers—but if you look closely you can still see the faint shimmer of a pattern in their fur. Leopards hunt and eat small and medium-sized animals, from hares to deer. Leopards are the only big cats that haul their kill up into a tree and wedge it between the branches to prevent other carnivores from stealing it.

The black rhinoceros has two horns. Others have only one.

People sometimes confuse hippos with rhinos. Perhaps this is because the two animals are often housed next to each other in the zoo, but apart from their almost hairless skin and their barrel-shaped bodies, they have very little in common. The rhinoceros is related to the horse. The hippo is not, in spite of its name, which means "river horse" in Greek. Instead, the hippo is related to the pig.

There are five types of rhinoceroses. None of them are very sociable. Apart from the black rhinoceros, most live their lives entirely alone. The horn of a rhinoceros is made from the same material as your hair and fingernails. The black rhinoceros shown here is the largest member of the rhinoceros family. It weighs up to 5,000 pounds (2,300 kg). Hippos can weigh even more—up to 8,000 pounds (3,600 kg).

The rhinoceros is an odd-toed ungulate, like the horse and the donkey. Odd-toed ungulates carry their weight mainly on their middle toe.

The hippo's eyes, nose, and ears are positioned so they remain above the water when the rest of its body is submerged.

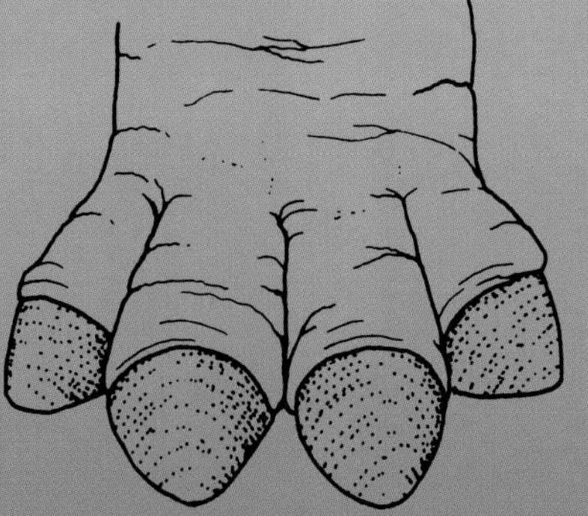

Hippos have four toes, so they are even-toed ungulates, like pigs and cows.

The hippopotamus spends some of its time on land, but is more at ease in the water. It likes calm water with dry, sandy banks for sunbathing. During the day it lazes half in and half out of the water. At night it comes out on land to feed, and can eat up to 88 pounds (40 kg) of grass each day. Hippos, like rhinos, are well armed against their enemies. A bull hippo's lower front teeth can be as long as 20 inches (50 cm). Hippos are sociable animals and herd together in groups of up to a hundred.
The hippopotamus has only one close relative, the pygmy hippo. It looks like a young hippo and weighs less than a tenth of its huge relative.

The head and neck of the European bison are not as massive and strong as the American bison's.

The European bison and the American bison are not only similar in appearance, but share the same terrible history—both have been hunted almost to extinction. Once, the European bison lived all across Europe and North Asia, but it was prized even by Stone Age hunters, as can be seen from many prehistoric cave paintings. As the largest European mammal, it was hunted more and more relentlessly until, 70 years ago, it was almost extinct. All the European bison alive today are descended from a handful of animals rescued and protected by Polish conservationists. European and American bison can interbreed, though of course this could never happen in the wild because they live on different continents.

In 15,000-year-old cave paintings you can see that prehistoric hunters stalked bison.

THE AMERICAN BISON OR BUFFALO

The American bison's back slopes down towards its hind legs.

This Native American painting shows the outline of a bison's head. Many North American Indian tribes once lived off the huge herds.

If you look carefully, you will notice a few small differences between the European bison and the buffalo. The North American bison carries its head lower. Its forequarters are more massive, and its hindquarters are smaller. An old bull will have much darker hair on his head and neck than on the rest of his body. A bull can be about a third heavier than a cow. With European bison, the difference between bull and cow is not so obvious.
Sixty million bison once lived on the North American prairies. When a herd stampeded, the earth thundered and quaked. When Europeans came, they shot thousands of these animals, often only for sport. Thanks to a last-minute conservation scheme, there are now 50,000 buffalo in the United States and Canada.

Grey seals belong to
the group known as earless seals.

Grey seals live around northern coasts. On land they move awkwardly because they cannot waddle along on their flippers like sea lions, but must wriggle on their bellies. They can, however, swim for longer distances, and dive deeper. They move through the water with powerful, sideways strokes of their tails and hindquarters. Grey seals like to live in calm water, and rest on sandbanks that are exposed by the ebb tide. The males and females are about the same size. The young males know how to swim as soon as they are born, and get milk from their mothers only for the first five weeks of their lives. By then they are almost four times as heavy as at birth.

An earless seal has no outer-ear structure. There is only an ear opening, which the seal can close when it dives.

California sea lions belong to the group known as eared seals. Males are more than three times heavier than the female shown here.

The sea lion has a small external ear.

You have almost certainly seen California sea lions in the zoo or at the circus. They are intelligent animals and can be trained to do many tricks. On land, sea lions are much more agile than seals. They draw their hind flippers in under their bodies and walk on all fours. When swimming, they move their flippers as if they were flying through the water.
In early summer California sea lions assemble on land in big herds. The males establish a territory and defend it day and night against rivals. Two weeks later the females, which are considerably smaller, come on shore, usually several females to each male. Each pregnant female gives birth to the pup that has been growing inside her since the last mating season. Once the young have been born, the sea lions mate again.

MOOSE

Moose look quite different from other deer. The bulbous nose and the beard are distinctive features.

The moose is the largest member of the deer family. Their long heads make them look rather like horses, but they are bigger—more than six and a half feet (200 cm) tall at the shoulders, and more than nine and a half feet (300 cm) from nose to tail. Most of their body hair is dark brown, though their legs are sometimes pale grey. Most older bulls have massive antlers with short branches. Like many deer, the moose sheds its antlers once a year. Then a new pair grows from the bony lumps on its forehead.

In Russia in the 1960s people tried to tame moose in the same way people have tamed reindeer. They were successful, and for the first time people rode on the backs of these huge animals.

A moose cow has no antlers at any time of year.

REINDEER OR CARIBOU

Reindeer reach only two thirds the height of a moose, and they weigh less than half as much. The males have large, finely branched antlers.

The female reindeer, unlike any other female deer, also has antlers.

The reindeer, like the moose, lives in the north lands, but reindeer are much smaller than moose. Their coats shade from brown through silver-grey to white, and the neck is usually paler than the rump.

Reindeer roam in large herds far across the tundra, eating grass and lichen, and since the Stone Age people have followed the herds and lived off them. No other deer travels over such vast distances. Many people of the north still live this way, though the animals are now also raised on farms.